Stellar Thoughts

星思维

第十二届“星艺杯”设计大赛获奖作品集

星艺装饰文化传媒中心　编著

暨南大学出版社
JINAN UNIVERSITY PRESS

中国·广州

图书在版编目（CIP）数据

星　思维. 第十二届“星艺杯”设计大赛获奖作品集 = Stellar Thoughts: Collection of Winning Works in the 12th “Xingyi Cup” Design Competition : 汉英对照 / 星艺装饰文化传媒中心编著. -- 广州 : 暨南大学出版社, 2024. 10. -- ISBN 978-7-5668-3990-9

Ⅰ. TU206; TU241.02

中国国家版本馆 CIP 数据核字第 2024HK0805 号

星　思维：第十二届“星艺杯”设计大赛获奖作品集
XING SIWEI: DI-SHIER JIE “XINGYI BEI” SHEJI DASAI HUOJIANG ZUOPINJI
编著者：星艺装饰文化传媒中心

出 版 人： 阳　翼
统　　筹： 杜小陆
责任编辑： 黄志波
责任校对： 刘舜怡　王燕丽
责任印制： 周一丹　郑玉婷

出版发行： 暨南大学出版社（511434）
电　　话： 总编室（8620）31105261
营销部（8620）37331682　37331689
传　　真：（8620）31105289（办公室）　37331684（营销部）
网　　址： http: //www.jnupress.com
排　　版： 广州良弓广告有限公司
印　　刷： 深圳市新联美术印刷有限公司
开　　本： 889mm × 1194mm　1/12
印　　张： 18
字　　数： 280 千
版　　次： 2024 年 10 月第 1 版
印　　次： 2024 年 10 月第 1 次
定　　价： 268.00 元

Create Happiness And Deliver Joy

设计幸福　播种快乐

星 艺 装 饰 文 化 传 媒 中 心

CONTENTS

1 住宅·工程实景作品

2 住宅·方案设计作品

3 大宅·工程实景作品

4 大宅·方案设计作品

5 公共·工程实景作品

6 公共·方案设计作品

Home Decoration Real-life Design Works

1 住宅·工程实景作品

归宅
COME BACK HOME

项目名称：归宅
项目设计：广东星艺装饰集团
设计师：郭宇

Project Name: Come Back Home
Program Constructed by: Guangdong Xingyi Decoration Group
Designer: Guo Yu

本案设计空间以灰、木、黑色调的强烈对比，呈现出现代简约的质感。设计师以纯粹直接的设计手法表达空间的大气与清爽的宜居感，为居住者打造一个可以回归心灵的安定居所。

本案设计营造出一个温暖而理性的空间，让家人们有更多互动沟通的共享空间，体验“出则繁华，入则宁静”的美好生活。

This case features a strong contrast of gray, wood, and black tones, presenting a modern minimalist aesthetic. The designer employs a straightforward approach to convey a sense of grandeur and refreshing livability, creating a serene haven for residents to return to their true selves.

A warm but rational space is successfully created, providing more room for family interaction and communication, embodying the beauty of "bustling outside, tranquil inside".

淳
PURE

项目名称：淳
项目设计：广东星艺装饰集团
设计师：王秋兰

Project Name: Pure
Program Constructed by: Guangdong Xingyi Decoration Group
Designer: Wang Qiulan

本案设计师与业主就地域文化和生活习惯进行了深入的交流，大胆取消了传统的餐厅区域，将餐饮空间与厨房完美结合，形成一个中岛台，原餐厅处布置了一张大长台，更便于创作。简洁高级的组柜，完美地收纳了日常琐碎之物，使得空间更加整洁。连廊的洄游动线，让卧室和手作工作室、客厅紧密相连，使得空间更加灵活舒适。在自然光下，材料的质感纹理更加凸显，让人感受到光影变化和时光变迁。

The designer discussed deeply with the owner on regional culture and life-style. They boldly eliminated the traditional dining area, perfectly integrating the dining space with the kitchen to form a central island. A large dining table is positioned on the original dining area, providing a more convenient environment for the owner's creativity. The simple and upscale storage cabinets efficiently organize daily stuffs, enhancing overall space cleanliness. The corridors connect the bedroom, studio, and living room in a smooth way, ensuring a flexible and comfortable living environment. In natural light, the texture and grain of materials are more pronounced, allowing people to perceive changes in light and the passage of time.

理想居所
IDEAL HOME

项目名称：理想居所
项目设计：广东星艺装饰集团
设计师：黎善寅、黄国展

Project Name: Ideal Home
Program Constructed by: Guangdong Xingyi Decoration Group
Designer: Li Shanyin, Huang Guozhan

本案是四口之家，夫妻二人育有一儿一女。业主特别注重对孩子的教育，希望在家中能有一个陪伴孩子成长的空间。小儿子喜欢在房子里肆意地奔跑、玩耍，大女儿经常练习弹钢琴，一家人都特别喜欢看书，希望有更多的储物空间，随时随地就能拿放书本。

为此，设计师以“家”为主题，打造了一个温暖、温馨的居住空间。在设计中，设计师把业主对生活的热爱和品质的追求注入空间的每一处细节。

This case is for a family of four, with a husband and wife raising a son and a daughter. The owners attach great importance to their children's education and require a space at home that supports children's growth. Their younger son enjoys running and playing freely around the house, while their elder daughter practices piano regularly. The family enjoys reading and desires ample storage space for books that can be accessed easily anytime.

To meet these needs, the designer take "home" as theme, creating a warm and cozy living environment. Throughout the design, the designer infused the owners' love for life and pursuit of quality into every detail.

揽月

EMBRACING MOON

项目名称：揽月
项目设计：广东星艺装饰集团
设计师：曾娟

Project Name: Embracing Moon
Program Constructed by: Guangdong Xingyi Decoration Group
Designer: Zeng Juan

家，是容纳漂泊灵魂的温馨港湾；家，是遮挡酷夏骄阳的如伞大树；家，是拂去疲惫的清凉雨丝。

当代室内装修已不再推崇复杂的装饰设计，极简主义越来越受青睐。本案设计简洁，没有杂乱的装饰，没有多余的色彩，极简设计给人一种无限的遐想。干净纯粹的美往往能直击人心。白色看起来并不起眼，有时或显单调，但其实白色是最接近艺术的颜色，永远不会过时，搭配上"少即是多"的极简设计理念，更能为空间设计带来无限的可能。柔软的模块组合沙发减弱了墙面的清冷，墙地一体的效果以及客厅大面积的落地窗，打造出明亮通透的视觉感受。

Home, is a warm harbor for wandering souls, a umbrella-like tree to prevent sun exposure, a rainfall to wash away fatigue.

Modern interior design has no preference to complicated decorations anymore while minimalism becomes more favorable. Without messy decorations and redundant colors, the simple design of this case leaves infinite room for imagination. The most pure beauty always hits hearts. White seems like a dull and overly simple color, but it actually aligns with an aesthetic sense. White is timeless, and along with the concept of "less is more", it brings unlimited possibilities to home design. Soft module-organized sofa diminished the coldness of the wall. Wall-floor integration effect and large floor-to-ceiling windows in the living room created a bright and transparent visual experience.

远·宇宙

REMOTE UNIVERSE

项目名称：远·宇宙
项目设计：广东星艺装饰集团
设计师：张超群

Project Name: Remote Universe
Program Constructed by: Guangdong Xingyi Decoration Group
Designer: Zhang Chaoqun

本案采用了经典的暗黑风格的设计，以深色调和复杂的纹理为主要元素，营造出一种不同于传统家装的全新空间体验。

灯光的运用是本案的亮点之一。照明设计旨在营造幽暗而又神秘的氛围。餐厅和起居室等聚会场所使用了柔和而不刺眼的灯光，还采用了多种材质来营造出舒适感和质感。木纹地板、金属灯具、皮质沙发等元素互相融合，营造出一种独特的视觉效果。

This case adopts a classic dark style, focusing on deep tones and complex textures to create a new spatial experience distinct from traditional home design.

The use of lighting is one of the highlights of this design. Lighting design aims to create a dim yet mysterious atmosphere. Soft and non-glaring lights are used in the dining room and living room, complemented by a variety of materials to enhance sense of comfort and quality. Elements such as wood grain flooring, metal lights, and leather sofas blend together to create a unique visual effect.

大玩家

BIG PLAYER

项目名称：大玩家
项目设计：广东星艺装饰集团
设计师：张羽

Project Name: Big Player
Program Constructed by: Guangdong Xingyi Decoration Group
Designer: Zhang Yu

本案是位于麓湖生态区的一套住宅，设计师用通透的白色与温润的木色，将柔光引入室内，营造出幽深与静谧的氛围。

入户即是客厅，木地板与柜体同色，归家就如同踏入森林秘境，入目即是一片清凉。整体户型较为规整，摒弃了传统客厅、餐厅、书房的设计，让孩子成为公共空间的“主人”。硬装则保留了清爽的原木风格，两侧光线流入，光影静默游走，时间如同静止了一般。

This case is a residential property located in the Luhu Ecological Area. The designer uses designing elements like transparent white and warm wood tones to introduce soft light into the interior, creating a tranquil and serene atmosphere.

Upon entering the living room, the seamless integration of wooden floors and cabinets in matching tones creates a cohesive and harmonious atmosphere. It evokes the feeling of stepping into a forest sanctuary, offering a refreshing and serene sight. The overall layout is rectangular, abandoning traditional designs like separate living rooms, dining rooms, and studies, making children the "masters" of the public space. The hard furnishing retains a refreshing natural wood style, with light streaming in from both side windows, and soft shadows quietly moving, giving the impression of time standing still.

等风来

WAITING FOR THE WIND

项目名称：等风来
项目设计：广东星艺装饰集团
设计师：周伟睿

Project Name: Waiting for the Wind
Program Constructed by: Guangdong Xingyi Decoration Group
Designer: Zhou Weirui

模糊边界，是为了建立起与自然的关系，从而接近设计的本质。大面积的落地玻璃模糊了建筑内外的界限，使室内室外形成一个自然有机的整体。将自然的变化引入室内，光于室内外穿梭，与自然相契合，唤醒生命的真正力量。室内以简单的材质色块划分出不同的功能区，结合不同的光源，又因光的无形，空间的区分变得无界，从而最大限度地释放空间。

Blurring the boundaries is aimed at establishing a connection with nature, thereby approaching the essence of design. Large floor-to-ceiling glass blur the line between the interior and exterior, creating a natural and intertwined whole. The design introduces the changes of nature indoors, allowing light to flow through both spaces. Light aligns with nature, awakening the true power of life. Inside, different functional areas are simply divided with materials, color blocks and lights. The intangible nature of light renders the spatial distinctions seamless, maximizing the sense of space.

禁止“焦绿”

DON'T BE ANXIOUS

项目名称：禁止“焦绿”
项目设计：广东星艺装饰集团
设计师：王超

Project Name: Don't be Anxious
Program Constructed by: Guangdong Xingyi Decoration Group
Designer: Wang Chao

本案营造了两种不同的氛围，在白天的时候，空间显得轻松愉快，而在夜晚，空间则显得高级优雅。色彩既是空间的点缀，也是空间气度的内在传达。本案选择一抹深邃黑绿作为空间叙事的基调，以情境述说关乎心灵的解读。灰色与绿色的神秘含蓄，奠定了沉稳内敛的空间基调，以弧线的造型优雅延展，穿透冷静肃穆的空间氛围。深邃之余，延展出几笔趣味盎然的清新绿意，浓淡相宜，鲜活灵动。

Two distinct atmospheres are created in this space: during the day, it feels light and enjoyable, while at night, it exudes elegance and sophistication. Colors decorate the room and deliver the inner quality of the room. Dark-green is chosen as the basic tone of the space, conveying an interpretation of the owner's heart. Grey and green, mysterious and humble, establish a calm and understated tone for the space. Curved shapes elegantly extend through the serene and solemn atmosphere, penetrating the space with a sense of tranquility. Against the deep background, some playful green color blocks are added, balancing in intensity and creating a vibrant, lively atmosphere.

M 府

M'S RESIDENCE

项目名称：M 府
项目设计：广东星艺装饰集团
设计师：梁柏诚

Project Name: M's Residence
Program Constructed by: Guangdong Xingyi Decoration Group
Designer: Liang Baicheng

建筑是一种思维方式，而非一门手艺。——勒·柯布西耶

我们规划的家也是思维方式的连接，只有思维产生了连接，情感空间才能得以升华。本案改造后，入户空间动线更加流畅，且采用“窥探”手法让各个空间产生交流。入户空间留出来的洞像是移动的画框，弧形的处理让空间灵动了几分，简洁低饱和的色彩也让人心安静了下来。

Architecture is a way of thinking in spite of a craftsmanship. —Le Corbusier

Similarly, our designed home is a connection of thoughts. Only when thoughts create connections can emotional spaces be sublimated. After the renovation, the entrance circulation becomes smoother and more convenient. The "peek-through" technique allows for interaction between spaces. The opening left in the entrance space resembles moving picture frames, with its curved shape adds a touch of fluidity to the space. The simple, low-saturation colors create a calming effect.

悠然·时和—翡翠城

YOURAN SHIHE：JADE CITY

项目名称：悠然·时和—翡翠城
项目设计：广东星艺装饰集团
设计师：晏敏芳

Project Name: Youran Shihe: Jade City
Program Constructed by: Guangdong Xingyi Decoration Group
Designer: Yan Minfang

本案业主对原户型的入户过道和主卫没有窗户极不满意，整体风格要求暗调、通透、时尚和雅致。本案将入户的墙体拆除一部分，加以一块线形艺术玻璃装饰，增加极窄鞋柜和灯光，并加宽进入次卧的过道。一进门便一眼看见半透客厅背景，瞬间增彩。

餐厅及过道的对景和谐呼应，搭配墨绿色彩，构筑了惬意的居家环境，餐桌餐椅以合围的形式营造了品享美食与情感互动交融的幸福场景，体块穿错的方式让原有结构不再过于独立地存在。

The owner was dissatisfied with the original layout's entrance corridor and master bathroom, because both of them lack windows. He requests a dark-toned, transparent, stylish, and elegant designing style. To address these issues, a part of the entrance wall was removed and replaced with a piece of linear art glass, with a narrow shoe cabinet and lighting added. The designer also widened the corridor leading to the secondary bedroom. Therefore, one can immediately see the semi-transparent living room upon entering, instantly enhancing the aesthetic appeal.

The dining area and corridor are harmoniously aligned, with blackish green color creating a cozy home environment. The dining table and chairs are arranged in a way that fosters both enjoying meals and emotional interaction, creating a scene of happiness. The interspersed functional parts transform the previously isolated structures, integrating into a cohesive whole.

沉浸
IMMERSION

项目名称：沉浸
项目设计：广东星艺装饰集团
设计师：姚迪

Project Name: Immersion
Program Constructed by: Guangdong Xingyi Decoration Group
Designer: Yao Di

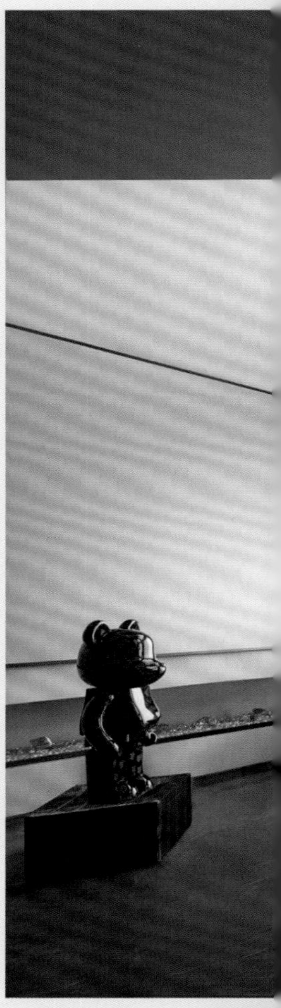

本案位于江景核心地带，开阔的视野格局符合居住者城市精英的身份气质，以及对简约疏朗且富有层次感的生活格调的追求。设计师将内敛沉稳、富有温暖情调的格局赋予空间，搭配精致的家居元素，在内敛与克制中，给人美而不张扬的视觉享受。

灰色大理石的温润气质与白色柜体的高贵质感和谐相融，成为空间的视觉焦点，艺术质感也被无限地放大。浅灰色磨砂皮质沙发点缀低饱和的质感色彩，木材、玻璃茶几更添典雅气质。整体去除繁复，进而营造出别样的高级感。安然的气息于空间内流动，勾勒出静谧的美好居所。空间亦灰而不冷、暖而有度，让床品、桌椅自身的质感与自然光线交融，延续简雅的调性，让情绪轻盈流淌，倦意得以释放，渲染自由之感。在生活与创意之间，生活的语言被完美地表达。在温润的色彩之间，与居住者共同寻找到精妙的共鸣。

The core location beside the river of this case offers a broad view that aligns with the owner's elite status and his request of high-quality life. A reserved and warm layout is infused into the space, complemented by exquisite furnishings, offering a refined but beautiful visual experience.

The warm quality of the gray marble seamlessly blends with the noble texture of the white cabinet, creating a visual focal point in the space and amplifying the artistic essence. The light-gray matte-leather sofa adds a touch of low-saturation color in the space, while the wooden and glass coffee tables further enhance the elegant atmosphere. By removing complexity, a unique sense of sophistication is created. Tranquil air flows through the space, showing a serene and beautiful home. The whole space is gray yet not cold, warm yet measured, allowing the textures of bedding and furniture to blend with natural light. This continuity of simplicity and elegance lets emotions flow lightly, alleviating fatigue and creating a sense of freedom. Between life and creativity, the language of living is perfectly expressed. Amidst the soothing colors, a subtle resonance with the residents is found together.

理想

IDEAL

项目名称：理想
项目设计：广东星艺装饰集团
设计师：黄马力

Project Name: Ideal
Program Constructed by: Guangdong Xingyi Decoration Group
Designer: Huang Mali

本案以“理想”为主题，将理想与现代生活相结合，以简洁明快的设计风格为主调，在总体布局方面尽量满足业主生活需求。主要装修材料为水泥灰饰面，以简洁的实木灰色线条及各种墙板隔断装饰，体现“理想”之感，营造温馨健康的家居环境。

This case, themed in "ideal", integrates idealism with modern living. The design features a clean and joyful aesthetic, with a layout meticulously planned to meet the owner's needs. The primary renovation materials is cement gray finishes, complemented by clean wood-made gray lines and various wall panel partitions. This design reflects the concept of "ideal", creating a warm and healthy home environment.

采桑子

CAI SANG ZI

项目名称：采桑子
项目设计：广东星艺装饰集团
设计师：张玲

Project Name: Cai Sang Zi
Program Constructed by: Guangdong Xingyi Decoration Group
Designer: Zhang Ling

本案是业主的第二套房子，依然选择由星艺装修，所以双方对彼此的了解更深入。前期和业主沟通方案时，男女主人的喜好是相悖的。女主人想要田园美式或乡村美式风格，但是考虑到男主人的接受度及本案整体空间尺寸，最后综合考虑，确定了现代美式的主体风格，硬装部分再添加一点浪漫法式线条点缀。

This case is the owner's second property, and they chose Xingyi Decoration Group once again, which has deepened the mutual understanding between both parties. During the initial discussions, the preferences of the husband and wife were quite divergent. The wife preferred a pastoral or country-style American design; while considering the husband's tastes and the overall space size, a modern American style was chosen as the primary theme. Additionally, some romantic French lines was added to the hard decoration to create a balanced design.

寒舍

MY HUMBLE HOUSE

项目名称：寒舍
项目设计：广东星艺装饰集团
设计师：何世荣

Project Name: My Humble House
Program Constructed by: Guangdong Xingyi Decoration Group
Designer: He Shirong

本案坐落在广西贺州爱莲湖畔，亲邻爱莲湖公园，远离城市的喧嚣，私享生活的快与慢、简与奢。为了实现家庭社交场景多元化，客餐厅选择了全开放式的一体化设计，布局灵动，轻松满足每个家庭成员的多层次需求，还融入些许坪庭小景的点缀，雅致而又素净，简约而不失错落有致。

This house is situated by Ailian Lake in Hezhou, Guangxi, adjacent to Ailian Lake Park. It is far away from the city's hustle and bustle, allowing its residents to privately enjoy the fast and slow, the simple and the luxurious aspects of life. To fulfill the need of diverse family life scenarios, the living and dining rooms are designed with an open, integrated layout. This flexible arrangement easily meets the varied needs of each family member. The design also incorporates small courtyard landscapes, adding an elegant and simple touch that is both minimal and layered.

浣花小筑

FLOWERY COTTAGE

项目名称：浣花小筑
项目设计：广东星艺装饰集团
设计师：张晓财、李帆

Project Name: Flowery Cottage
Program Constructed by: Guangdong Xingyi Decoration Group
Designer: Zhang Xiaocai, Li Fan

住宅和家庭的关系是一种适配并兼容拓展的关系，住宅容纳着家庭成员的日常活动，且将随着时间延伸到未来的生活。在这栋六层联排别墅中，设计师准确地平衡居住空间承载的日常性与互动性，依照不同空间的使用形态进行逐级分层，明确每一层的空间场域与氛围。

The relationship between residence and family is a kind of adaptation and expansion. The residence accommodates the daily activities of family members while extending into the future. In this six-story villa, the designer has skillfully balanced the daily needs and interactions required by the family. The design implements a tiered layout according to the different usages of each space, clearly defining the ambiance and function of every floor.

Company
LOOKBOOK
CLOTHING

星美

STARRY BEAUTY

项目名称：星美
项目设计：广东星艺装饰集团
设计师：马婷婷

Project Name: Starry Beauty
Program Constructed by: Guangdong Xingyi Decoration Group
Designer: Ma Tingting

本案整体的设计风格定位为现代极简风，辅以意式简约与轻奢风的元素特征，采用无主灯的设计，大胆运用的色彩、材料、灯光都在设计中扮演着重要的角色。色彩以暖色调为主，通过材料和质感的搭配，营造出简约、大气、优雅的空间感。同时，注重光线的运用，充分利用自然光线，让空间更加通透明亮。

The overall design style of this project is positioned as modern minimalist, complemented by elements of Italian simplicity and light luxury. It features a design without a central light, boldly incorporating colors, materials, and lighting that play significant roles in the design. The color scheme is predominantly warm tones, complemented by the combination of materials and textures to create a sense of simplicity, spaciousness, and elegance in the space. At the same time, emphasis is placed on the use of light, making full use of natural light to create a more transparent and bright space.

Home Decoration Scheme Design Works

2 住宅·方案设计作品

徐宅

XU'S RESIDENCE

项目名称：徐宅
项目设计：广东星艺装饰集团
设计师：张杰

Project Name: Xu's Residence
Program Constructed by: Guangdong Xingyi Decoration Group
Designer: Zhang Jie

在本案大平层的空间设计中，设计师努力挖掘更多的趣味性和可能性。

简约的设计让空间拥有了更多的延展性。本案以业主理想中的生活状态为设计出发点，重新构建空间秩序，打造更为自然的生活动线。

In this case, the designer explores more interesting and possible aspects within the expansive flat.

The minimal design allows the space to have more extension and potential. By starting from the ideal living state of the residents, the designer reconfigures the spatial order to create the most natural living flow for the family.

春晓

SPRING MORNING

项目名称：春晓
项目设计：广东星艺装饰集团
设计师：刘振

Project Name: Spring Morning
Program Constructed by: Guangdong Xingyi Decoration Group
Designer: Liu Zhen

一个珍贵而理性的图式一朝被实录下来，将永不消逝。即便我们离去很久，它仍将具有旺盛的生命力，并愈发坚持地展示自己。

A precious and rational schema once actualized will never vanish. Even though have been away for a long time, it will exhibit vitality and persistently present itself.

光·璞

GUANG PU

项目名称：光·璞
项目设计：广东星艺装饰集团
设计师：张晓财、李帆

Project Name: Guang Pu
Program Constructed by: Guangdong Xingyi Decoration Group
Designer: Zhang Xiaocai, Li Fan

本案中，设计师通过光线的明暗对比和深浅变化来表现空间的层次感。

整体以引景、置景入室的思路翻开空间序章，透过光线与建筑量体动线的安排，移步换景，感受“处处皆生活”的惬意。以庭院为轴线贯穿起居室和餐厅中西厨，使室外景观与多元互动的社交空间相互呼应。留白空间用深色石材与金属构件作画，带来以建筑形体张力与光影为感知的景观体验，仿佛置身于充满自然气息的情景互动社交空间。自然景观引入是空间氛围重点，生活机能与光影景观实现情景化，带来阳光景观无处不在的自然惬意。

In this case, the designer uses contrasts of light and shadow, along with variations in depth, to convey the layers of space.

The overall design concept places the scenery in the forefront of the interior space. Through the arrangement of light and circulation, each step of visitor reveals a new scene, embodying the ease of "life is everywhere". The courtyard serves as the axis, connecting the living room and the dining area with both Western and Chinese kitchens, creating a corresponding relationship between the outdoor landscape and the diverse social spaces. The blank spaces are filled with dark stone and metal, offering a visual experience where architectural tension coexists with light and shadow. This creates an interactive social space that makes people feel immersed in nature. Introducing natural landscapes is crucial for creating the spatial ambiance. The living areas are seamlessly integrated with light and shadow, ensuring that sunlight and scenic views are ever-present, bringing a sense of natural ease and comfort to the home.

落地窗环绕外景，连接客厅、餐厅、走廊与卧室区域，全面打开视野，完整呈现一线江景。中国书画讲究“留白”，也许就是西方常说的“少即是多”。形象化的结束往往是意象化的开始。黑色墙面犹如深邃夜空，聚焦目光。

Floor-to-ceiling windows surround the exterior view, connecting the living room, dining room, hallway, and bedrooms, fully presenting an uninterrupted river view. Chinese painting emphasizes "negative space", which may be akin to the Western concept of "less is more". When the design is taken to its utmost perfection, abstraction begins. The black walls, resembling a deep night sky, draw and focus attention.

折和间

ZHE-HE-JIAN

项目名称：折和间
项目设计：广东星艺装饰集团
设计师：刘诚

Project Name: Zhe-He-Jian
Program Constructed by: Guangdong Xingyi Decoration Group
Designer: Liu Cheng

折：“折”即“折中”，是中华传统文化的一个组成部分。一折一合，一遮一示，尽显朴实、儒雅、含蓄。

和：“和”是中国传统哲学的重要范畴。老子说：“冲气以为和。”荀子说：“万物各得其和以生。”“和”揭示了宇宙运动规律，是自然的最佳境界和终极状态。

间：东方的“间”独立于周围空间，强调两者之间肉眼看不见的关系，东方文化中的关系性使“空白”具有意义。

Zhe: "Zhe" means "balance", a key concept in traditional Chinese culture. Fold and unfold, conceal and reveal, in this process simplicity, elegance, and subtlety all fully displayed.

He: "He" is an important category in Chinese philosophy. Laozi said, "Harmony is created by balancing the flow of energy." Xunzi said, "All things thrive when they achieve their proper harmony." "Harmony" reveals the laws of cosmic movement and represents the highest realm and ultimate state of nature.

Jian: "Jian" in oriental culture refers to the surroundings that something is isolated from, emphasizing the invisible relationships between them. This relational aspect imparts meaning to emptiness.

曲中平直

STRAIGHTNESS IN THE CURVES

项目名称：曲中平直
项目设计：广东星艺装饰集团
设计师：吴坡

Project Name: Straightness in the Curves
Program Constructed by: Guangdong Xingyi Decoration Group
Designer: Wu Po

本案将传统与现代相结合，设计风格简洁明快，从简单中追求艺术美，使设计变得简约而不简单。客厅作为家庭生活区域之一，既是娱乐、休闲、聚餐的活动场所，又是接待客人的社交场所。色彩以咖啡色为主，稳重大气，品位高雅。居室整体采用中性色调，稳重大气却又不失品位。墙面都采用白色，地板都选用木质，居室设计的核心是床和衣橱，其他家具的摆放根据业主自己的习惯来设计。厨房是制作美食的场所，考虑了干净、通风和使用方便等诸多因素，颜色的选择以清洁、卫生为主，就餐环境的营造和整个居室风格保持协调一致。

This project combines traditional and modern elements, featuring a design style that is clean and bright. It seeks artistic beauty in simplicity, making the design minimal different from ordinary. The living room, as one of the family's main living areas, serves as a space for entertainment, relaxation, and dining, as well as a social venue for hosting guests. The living room is dominated by coffee tones, while the entire house features neutral hues; both places blend stability and grandeur with refined elegance. The walls are painted white, and the floors are wooden. The core elements in the bedroom are the bed and wardrobe, while the arrangement of other furniture is based on the owner's personal preferences. The kitchen, designed as a space for preparing delicious meals, prioritizes cleanliness, ventilation, and ease of use. The color scheme focuses on cleanliness and health, while the dining room is crafted to co-ordinate with the overall style of the home.

半室·半园

HALF ROOM-HALF GARDEN

项目名称：半室·半园
项目设计：广东星艺装饰集团
设计师：王树涛

Project Name: Half Room-Half Garden
Program Constructed by: Guangdong Xingyi Decoration Group
Designer: Wang Shutao

本案远离城市中心，回归乡村，可以让人感受到生活中的“真”。原木和钢材的拼接在空间中有一种结构力学的关系。简洁留白的立面与精致细腻的材质建立起空间的室内外关系，让人寄情于自然之中，享受美好的时光。

This project is situated away from the city center, returning to a more rustic environment, allowing people to experience the sense of "true" in life. The combination of raw wood and steel creates a structural dynamic. The simple, unadorned surfaces and the refined, delicate materials establish a connection of the interior and exterior, allowing people to immerse themselves in nature and enjoy beautiful moments.

自然而然

NATURALLY

项目名称：自然而然
项目设计：广东星艺装饰集团
设计师：王树涛、吴静雯

Project Name: Naturally
Program Constructed by: Guangdong Xingyi Decoration Group
Designer: Wang Shutao, Wu Jingwen

经过精心设计、修饰的面层占据着日常的主流审美，“结构即装饰”的理念渐渐变得清晰起来。例如地下室的光线梳理，设计师让楼顶的自然光通过天井倾泻到地下室区域。一楼整个客厅改变原有墙体结构，客餐厅区域连贯一体，正面落地玻璃使室内与室外和谐共生，业主可以打开心扉，与自然对话。框景的轮廓隔断丰富了空间构成，灵动的动线设计有效拓展了空间的纵深度。

Careful and refined designs now represent the mainstream aesthetic, with the concept of "creating structure is decoration" becoming increasingly clear. For example, in the basement's design, the designer prefers to introduce natural light from the roof, allowing it to cascade through the skylight into the basement area. On the first floor, the original wall structure has been modified to create a seamless integration of the living and dining areas. Floor-to-ceiling glass on the facade fosters a harmonious connection between indoor and outdoor spaces, allowing residents to open up and engage with nature. Partitions that incorporate landscaping enrich the spatial composition, while the dynamic circulation design effectively extends the depth of the space.

理白

MARBLE WHITE

项目名称：理白
项目设计：广东星艺装饰集团
设计师：李鲜红

Project Name: Marble White
Program Constructed by: Guangdong Xingyi Decoration Group
Designer: Li Xianhong

拒绝堆砌，让空间留白，用最少的设计赋予空间无限想象，也令空间有了更大的包容力，营造舒适的居家环境，让业主在这里可以完全展示自己的生活方式与态度。

The designer leaves the space with endless possibilities instead of clutter, also enhances the space's inclusiveness. Residents can fully showcase their lifestyle and attitude in this room, in the same time they can create a comfortable home environment.

低调奢宅

DISCREET LUXURIOUS RESIDENCE

项目名称：低调奢宅
项目设计：广东星艺装饰集团
设计师：潘建平

Project Name: Discreet Luxurious Residence
Program Constructed by: Guangdong Xingyi Decoration Group
Designer: Pan Jianping

项目原建面积为 146 ㎡，室内大多是中空结构，为满足业主居住要求，室内部分中空加建楼板，最终套内面积扩大至 360 ㎡。本案以直线条为主，利落简洁的风格深入人心，使人与环境、室内与室外花园彼此和谐共存，排斥一切杂乱，化繁为简，设计变得简约而不简单。

The original building area of the project was 146 square meters, among which most are of hollow structure. To meet the owner's requirements, additional floor were added in part of void area, resulting in a final usable area of 360 square meters. The design features a sleek and minimalist style with clean lines. It harmoniously integrates indoor room and outdoor gardens, rejecting clutter, simplifying complexity, and making the design simple but not simplistic.

常念
MISSING OFTEN

项目名称：常念
项目设计：广东星艺装饰集团
设计师：卢兴

Project Name: Missing Often
Program Constructed by: Guangdong Xingyi Decoration Group
Designer: Lu Xing

本案业主是一位十分优雅、温和的女性，由于远嫁外地，牵挂父母，于是选择渭南作为父母的养老之所。她喜欢修身养性，对生活的态度亦是以质衡量，唯愿岁月安暖。

进入室内，整个空间使用暖色涂料，抛弃了繁杂的装饰，大面积留白，在干净温润的底色中，亚麻与实木的材质被烘托得更加有温度。家具虽然不多，但是精致无比，宽敞的客厅和简洁的装饰相得益彰，家具摆放显得自然而又朴实、静谧而又舒适。

The homeowner in this case is a very elegant and gentle woman, who living far from home, cares deeply for her parents. She has chosen Weinan as a place for her parents' retirement. She enjoys cultivating herself and values life quality. All she desires is a beautiful life.

Upon entering, the entire interior space is painted in warm colors, abandoning elaborate decorations and leaving large areas of blank space. Against this clean and gentle backdrop, the textures of linen and solid wood are highlighted, representing a warm atmosphere. Although the furniture is minimal, it is exceptionally refined. The spacious living room complements the simple decorations, creating a natural, unpretentious, tranquil, and comfortable atmosphere.

谢宅

XIE'S RESIDENCE

项目名称：谢宅
项目设计：广东星艺装饰集团
设计师：潘远石

Project Name: Xie's Residence
Program Constructed by: Guangdong Xingyi Decoration Group
Designer: Pan Yuanshi

侘寂与时间的对抗，取决于事物的本质，如果一开始就是繁华绚烂，那么终将折服于时间，归于平淡。在某种意义上，侘寂是对抗时间的一种神奇之力，从寻常事物中寻到不寻常的美，朴素、寂静、自然，如此甚好。

The struggle between wabi-sabi and time depends on the essence of things. Though something may begin with splendor and grandeur, it will eventually surrender to time and return to simplicity. In a sense, wabi-sabi represents a magical force against time, discovering extraordinary beauty in ordinary things—simplicity, serenity, and naturalness. So beautiful in its own way.

简而精致
SIMPLE YET REFINED

项目名称：简而精致
项目设计：广东星艺装饰集团
设计师：李磊

Project Name: Simple yet Refined
Program Constructed by: Guangdong Xingyi Decoration Group
Designer: Li Lei

本案以简约而细腻的笔触，为业主塑造了一个现代风的家，拥有当下设计的舒适便捷，却没有快节奏的氛围。一切都以业主的心境为出发点，创造适合的功能，以精巧的细节赋予空间仪式感，让生活变得简单而又精彩。

科技联动人与世界，却也让人在复杂纷繁里迷失。而家可以剥离这些复杂的关系，令人回归自我。小书架代替电视，开辟一方寻找自我的独处空间；沙发相对而立，舒适宽绰，足够一家人对坐。

With a simple but refined touch, this case shapes a modern-style home for the residents by offering the comfort and convenience of contemporary design while rejects fast-paced design trends. Everything is designed for the residents' state of mind, creating functional spaces and imbuing them with details, making life simple but splendid.

Technology connects people with the world but can also make people lost in complexity. A home, however, can strip away these complexities and bring one back to himself. A small bookshelf replaces the television, carving out a space for solitary introspection. Two sofas face each other, which is comfortably, spacious enough for the whole family to gather and sit together.

谦虚与傲慢
HUMBLE AND PRIDE

项目名称：谦虚与傲慢
项目设计：广东星艺装饰集团
设计师：郑桂鑫

Project Name: Humble and Pride
Program Constructed by: Guangdong Xingyi Decoration Group
Designer: Zheng Guixin

慵懒、随性却神秘莫测的黑白灰，将精致极简、从容优雅藏于内，将平静内敛呈现在眼前。

This house is of leisure, casual yet enigmatically black, white, and gray. It hides exquisite minimalism and graceful composure within, presenting calm and restraint right before your eyes.

静屿

LONELY ISLAND

项目名称：静屿
项目设计：广东星艺装饰集团
设计师：陈健劲

Project Name: Lonely Island
Program Constructed by: Guangdong Xingyi Decoration Group
Designer: Chen Jianjin

本案是一个极简之家，开放的公共空间犹如一个巨大的容器，洄游动线将起居室、多功能区、餐厨空间串联起来，赋予整个居所灵动的活力，自然气韵通透流畅，自由流淌。为了让空间看起来更为纯粹，浑然一体，内在又散发出家的温馨，整体以木色为主色调，部分点缀白色及黑色。

This project is a minimalist home, where the open public space resembles a vast container. A circulation connects the living room, multifunctional area, and kitchen-dining space, endowing the entire residence with dynamic vitality. Natural air is flowing throughout smoothly and freely. To create a more cohesive and purer look that exudes the warmth of a home, the primary color scheme is wood tones, with white and black interspersed.

时序

TIME ORDER

项目名称：时序
项目设计：广东星艺装饰集团
设计师：周家兴

Project Name: Time Order
Program Constructed by: Guangdong Xingyi Decoration Group
Designer: Zhou Jiaxing

本案融合业主对理想生活的期许，运用利落的线条与张弛有度的空间节奏，搭配质感满满的装饰，在美学与舒适之间取得平衡，构筑优雅的品质空间。以契合现代的生活态度为核心，整体造型简洁，摒弃一切多余元素，通过细节材质的变化陈述空间，表达一种不流于大同的审美意趣。

This case merges the residents' expectations of an ideal life into the design, using sharp lines and a balanced spatial rhythm, coupled with rich decorative textures, to achieve a balance between aesthetics and comfort, and build an elegant and high-quality space. With a modern attitude towards life as the core, the overall design is simple, discarding any redundant elements, and expressing the space through the details of material changes, conveying a unique aesthetic interest.

觅·境

SEEKING REALM

项目名称：觅·境
项目设计：广东星艺装饰集团
设计师：石全刚

Project Name: Seeking Realm
Program Constructed by: Guangdong Xingyi Decoration Group
Designer: Shi Quangang

本案打破固有思维，在满足刚性需求功能外，通过家具的摆放，将人的视线聚焦过来。再把动线做成最便捷的环形，为家人创造更多的亲子互动空间。

This project breaks away from conventional thinking by not only meeting rigid functional requirements but also focusing people's attention through the arrangement of furniture. It establishes the most convenient circular flow to create more interactive spaces for family members, especially enhancing parent-child interactions.

Grand House Real-life Design Works

3 大宅·工程实景作品

栖园·静谧

SERENE HAVEN

项目名称：栖园 · 静谧
项目设计：广东星艺装饰集团
设计师：文泽林

Project Name: Serene Haven
Program Constructed by: Guangdong Xingyi Decoration Group
Designer: Wen Zelin

万物将被一次又一次地改变，唯一不确定的是被谁改变。在繁华都市里选择一个宁静之处，借四季的变化演绎空间不同的情感。

Everything will be changed again and again, the only uncertainty is by whom they will be changed. In a bustling city, choose a tranquil place to call home. As the seasons change, the emotions conveyed by the space also vary.

生活，自有时光相伴；家居，更需品位作答。本案将自然之美引入室内，营造有生命力的空间氛围。高级不是流于表面的美，而是深入内里的精致，以高品质人居新范式造就独特的私享空间。

家的归属感是目光所及的温暖。用自然之色慢调自然而然的生活。爱这份秩序井然，更爱这份仪式感。

Life is accompanied by time; while home decoration echoes owner's taste. This case introduces the natural landscape indoors, creating a vibrant spatial atmosphere. True sophistication reflects by deep-rooted elegance in spite of superficial beauty. The design establishes a new paradigm of high-quality living and creates a unique private haven.

The sense of belonging in home is reflected in the warm details that is visible everywhere. Use the colors of nature to subtly orchestrate a natural lifestyle. Love this orderly harmony, and love this sense of ritual more.

佛奥湾

FO'AO BAY

项目名称：佛奥湾
项目设计：广东星艺装饰集团
设计师：余文文、骆家骏

Project Name: Fo'ao Bay
Program Constructed by: Guangdong Xingyi Decoration Group
Designer: Yu Wenwen, Luo Jiajun

家，是心之所向，也是情感的交融之所。

家是一面生活的镜子，或是慌忙杂乱，或是安定沉静；或是趋向物化，或是趋向精神。

家应该是包含空间、色彩、情感的多维度立体感受的精神家园，将自身融入其中，感受细腻入微的材质、肌理、构造之间的参差质感，细查光影在不同时刻投射的几何形状，寻找人与空间最恰当的相处方式。

Home, is where your heart belongs to and where emotions interweave.

Home is a mirror of your life: it may be busy or peaceful; it can lean towards materialism or spirituality.

A home should be a spiritual sanctuary that encompasses a multidimensional sensory experience of space, color, and emotion. It invites you to immerse yourself in it, feeling the subtle textures, materials, and structures. At home, you may also observe the shapes cast by light and shadow at different times, and seek the most harmonious way to interact with it.

合·和

UNITY AND HARMONY

项目名称：合·和
项目设计：广东星艺装饰集团
设计师：万丹

Project Name: Unity and Harmony
Program Constructed by: Guangdong Xingyi Decoration Group
Designer: Wan Dan

“合”，意为连接、融合、结合、综合，代表了空间功能的融合、所用材料的结合，以及设计效果的综合。“和”，意为设计师与业主之间的关系。

本案以灰色为主色调，在定制产品和造型上以金属的深灰色、黑色为点缀，打造一个有层次感的空间。浴室外的阳台上特意设置了植物小景，在阳光与风的作用下增添了生活中的幸福感。设计是关注人、空间本质和品质细节的。本案在自然与精致之间取得平衡，强调协调与融合，希望能让业主感受到生活中的温暖与幸福。

"He (Unity)", means connection, integration, combination, and synthesis. It represents the fusion of spatial functions, the combination of materials, and the comprehensive effect of design. "Harmony" means the relationship between the designer and the homeowner.

This case uses gray as the main color, with deep gray and black metal accents in custom products and designs, creating a layered space. A small plant is specially set up on the balcony outside the bathroom, enhancing the sense of happiness through the influence of sunlight and breeze. Design focuses on the essence of homeowner, space itself, and the details. This case strikes a balance between simplicity and complexity, emphasizing harmony and integration, aiming to make the homeowner feel the warmth and happiness of life.

法合

FRENCH-STYLE COMBINATION

项目名称：法合
项目设计：广东星艺装饰集团
设计师：辛建华

Project Name: French-style Combination
Program Constructed by: Guangdong Xingyi Decoration Group
Designer: Xin Jianhua

本案是一栋三层法式组合院别墅，分地上两层和地下一层，占地面积约为 260 m²。整体设计为舒适、豪华和传统的美式风格。色彩以中性色为主，有米白色、米黄色、浅灰色等，营造出温暖和舒适的氛围。本案通过合理的平面布局，以及灯光、家具、墙面的装饰细节，动静结合，打造出一个充满美式风情的室内空间。

This project features a three-floor French-style combined villa, comprising two above-ground floors and one basement level, covering an area of approximately 260 square meters. Its overall design emphasizes comfort, luxury, and traditional American style. The color predominantly consists of neutral tones such as off-white, beige, and light gray, creating a warm and cozy atmosphere. Through a well-planned layout, lighting, furniture, and wall decorations, the designer seamlessly blends dynamic and tranquil elements to create a space brimming with American-style charm.

慢·时光

SLOW TIME

项目名称：慢·时光
项目设计：广东星艺装饰集团
设计师：罗斯梅

Project Name: Slow Time
Program Constructed by: Guangdong Xingyi Decoration Group
Designer: Luo Simei

全屋奶咖色与胡桃木相得益彰，沉稳大气，温暖宁静。潘多拉岩板遇上沉默的黑色，总能产生视觉的冲击与空间的平衡。在暖意融融的奶咖色空间里，用橙色的单椅点缀出一抹艳丽。阳光透过梦幻的褶皱，空间静下来，时光慢下来，心也随之安静。

The entire home features a harmonious blend of creamy coffee tones and walnut wood, creating an elegant, warm and tranquil atmosphere. Pandora stone slabs paired with silent black evoke a striking visual impact and balance the spaces. In this warm, creamy coffee-colored space, an orange accent chair adds a vibrant touch. Sunlight filters through dreamy folds, calming the space, slowing down time, and bringing a sense of peace to the heart.

黑与光的对话

DIALOGUE OF DARK AND LIGHT

项目名称：黑与光的对话
项目设计：广东星艺装饰集团
设计师：姚国健

Project Name: Dialogue of Dark and Light
Program Constructed by: Guangdong Xingyi Decoration Group
Designer: Yao Guojian

在原始住宅结构中，中规中矩的柱子和墙体结构不仅使室内与室外自然景致产生隔阂，还使各个空间缺乏联系。设计师重新调整了布局与动线，在各个空间中寻找彼此之间的关联，有意识的规划使得整体结构变得清晰，为业主带来了一个全新的视角。 设计师用沉稳的黑色搭配大面积铺陈的温润木质地板，为业主营造出舒适的起居环境，低调中尽显优雅气质。

In the original structure, the conventional columns and walls not only created a barrier between indoor and outdoor views but also resulted in disconnection between spaces. The designer reconfigured the layout and circulation, seeking connections between the various areas. This intentional planning clarified the overall structure, providing the homeowner with a fresh perspective. A refined black palette paired with extensive warm wooden flooring were applied to create a comfortable living environment, featuring subtle and elegant ambiance.

春日偶成
SPRINGTIME SERENDIPITY

项目名称：春日偶成
项目设计：广东星艺装饰集团
设计师：张海涛

Project Name: Springtime Serendipity
Program Constructed by: Guangdong Xingyi Decoration Group
Designer: Zhang Haitao

家作为灵魂的栖息之所，表达的是一种立场，也是一种态度。在业主原有的诉求基础上，本案增添了更多的色彩、线条、功能、布局等，打造出一个安静淡然、清新素雅的居住空间。

与自然同居，与山月同歌，是现代都市人向往的写意生活。屏风造型的床铺背景，用灰调梅花和山水刺绣作映衬，尘世繁华间隐匿着宁静祥和，也为安逸和谐的空间增添了几分艺术的从容。

Home, as a sanctuary for the soul, represents both a stance and an attitude of the homeowner. Building on the homeowner's requests, this project incorporates additional colors, lines, functions, and layouts to create a serene, understated, and refreshingly elegant living space.

Living in harmony with nature and singing along with the mountains and moon represent the idealized lifestyle that modern citizens yearn for. The bed's backdrop features a screen-like design, complemented by gray-toned plum blossom and landscape embroidery. Amidst the splendor, it subtly conceals tranquility, adding an artful touch of composure to the serene space.

美湖私宅

BEAUTIFUL LAKE PRIVATE RESIDENCE

项目名称：美湖私宅
项目设计：广东星艺装饰集团
设计师：朱小宁

Project Name: Beautiful Lake Private Residence
Program Constructed by: Guangdong Xingyi Decoration Group
Designer: Zhu Xiaoning

本案业主希望打造一个安静、大气、内敛的居家环境，经过几次深入沟通，最终定位为意式轻奢风格。本案基于业主的生活习惯与需求，对空间重新进行规划，注重对功能化布局的统筹把控。此外，合理设置多动线，凸显整体流动性。同时融合审美情趣，以克制纯粹的设计语言，演绎非凡品位与格调。

The owners of this project desired to create a serene, grand, and reserved living environment. After several in-depth discussions, we settled on a modern Italian luxury style. Based on the owners' lifestyle and needs, we rearranged the space, focusing on a functional layout that ensures cohesive control. Additionally, we strategically incorporated multiple circulations to enhance overall fluidity. Meanwhile, we blended aesthetic charm with a restrained and pure design language. Finally, we achieved an extraordinary sense of taste and sophistication.

剑桥世界里

JIANQIAO SHIJIELI

项目名称：剑桥世界里
项目设计：广东星艺装饰集团
设计师：帅伯尤

Project Name: Jianqiao Shijieli
Program Constructed by: Guangdong Xingyi Decoration Group
Designer: Shuai Boyou

本案意图营造一个层次丰富、内容清爽的美式空间。淡雅的空间布景，搭配一些藏蓝青，空间马上就有了稳重感。这是悠然肆意里的沁人心脾，也是暖意温存下的最初体验，显现出源源不断的活力与生机。本案利用色彩的层次对比，展现出居室的高贵典雅，适宜的装饰凸显出豪华浪漫的格调。

In this case, the designer aims to create an American-style space that is rich in layers and refreshingly clear in furniture. A subtle and elegant space decoration, paired with touches of deep navy colors, immediately imbues the space with a sense of stability. It's like a touch of coolness in a leisurely life and also like a warm initial experience, revealing an ongoing sense of vitality and energy. This case uses layered color contrasts to represent the home's noble elegance, while appropriate decorations highlight its luxurious and romantic style.

荏苒岁月

TIME GOES BY

项目名称：荏苒岁月
项目设计：广东星艺装饰集团
设计师：叶翔

Project Name: Time Goes by
Program Constructed by: Guangdong Xingyi Decoration Group
Designer: Ye Xiang

对于祖宅重建，设计时要考虑的因素很多，既要“面子”，也要“里子”，最重要的一点是要有可传承的精神气韵。本案最终将室内风格定位为新中式，用现代的手法和材质将中式的气韵含蓄地表达出来。

大气挑空的客厅，白色水墨的岩板配上温润的实木板材，彰显了应有的气场，又不会显得张扬，一切都恰到好处。茶室将花园景观自然过渡，再沏上一壶好茶，便是人间好时节。

When rebuilding a family home, many factors must be considered, balancing both aesthetics and functions. Most importantly, the design should embody a spirit of inheriting traditions. This case ultimately embraces a modern Chinese style, subtly expressing traditional Chinese charm through contemporary methods and materials.

The spacious living room features a high ceiling, adorned with white ink-patterned stone slabs and warm wooden panels. It strikes the perfect balance between presence and restraint. The transition from the tea room to the garden is seamless. Brewing a pot of fine tea in the tea room turns it into a perfect moment of tranquility.

冠江·远香亭

GUANJIANG YUANXIANG TING

项目名称：冠江·远香亭
项目设计：广东星艺装饰集团
设计师：陶佳彤

Project Name: Guangjiang Yuanxiang Ting
Program Constructed by: Guangdong Xingyi Decoration Group
Designer: Tao Jiatong

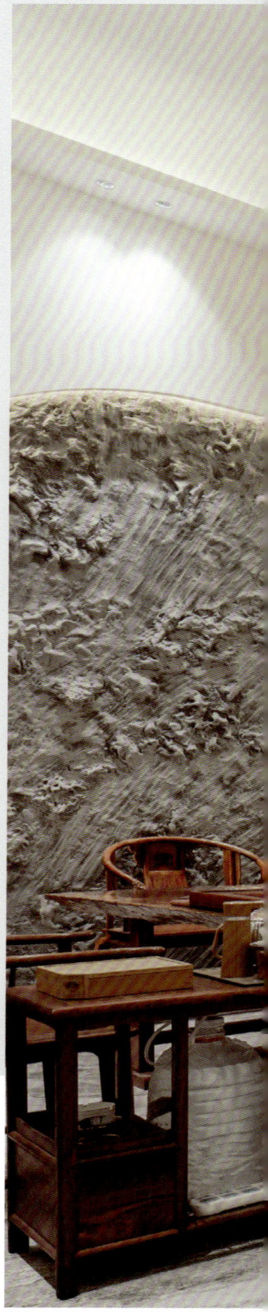

本案整体以灰色调山水连纹大理石贯穿整座房子。一楼的挑高客厅背景用白色哑光岩板营造出简约的氛围。公共区域用哑光拉丝不锈钢门套作点缀，给整个家增添了一丝高级感。喝茶区域用传统中式花格造型，中间用中国圆灯光造型烘托出古香古色的茶文化氛围。园林入口用迎客松和小白色鹅卵石堆营造休闲又典雅的园林景观。整个方案既简约又不失中式韵味。

Gray-toned marble with continuous landscape-like patterns is a design element that runs throughout the entire house. In the high-ceilinged living room on the first floor, white matte stone slabs are used for the backdrop, creating a minimalist atmosphere. The public areas are decorated with matte brushed stainless-steel door frames, adding a sense of sophistication to the entire home. The tea area features traditional Chinese flower-like pattern designs, with a central round Chinese lighting that enhances the ambiance of tea culture. The garden entrance is decorated with a welcoming pine and small white pebbles, creating a relaxing and elegant landscape. The entire design is full of minimalist and Chinese-style charm.

回归
RETURN

项目名称：回归
项目设计：广东星艺装饰集团
设计师：周兵、陶佳彤

Project Name: Return
Program Constructed by: Guangdong Xingyi Decoration Group
Designer: Zhou Bing, Tao Jiatong

本案主要使用了岩板、不锈钢、艺术玻璃、大理石等现代时尚的材质。从负二楼到一楼的公共区域主要是用于会客接待。整个空间用的是灰色调的连纹岩板，墙面用白色岩板营造简约的氛围。负二楼和一楼客厅都做了挑高层，超高的背景墙搭配着山水图文的岩板，再配上新中式的家具和富有中式韵味的饰品，尽显会客厅的大气。

This project primarily uses modern and stylish materials such as stone slabs, stainless steel, art glass, and marble. The public areas from the second basement to the first floor are primarily used for hosting and entertaining guests. The entire space features gray-toned continuous pattern stone slabs, with white stone slabs on the walls to create a minimalist atmosphere. Both the second basement and first-floor living rooms have high ceilings, with an impressive backdrop of stone slabs featuring landscape imagery. These are complemented by modern Chinese furniture and accessories with traditional Chinese charm, highlighting the grandeur of the reception area.

庐境

LU JING

项目名称：庐境
项目设计：广东星艺装饰集团
设计师：马婷婷

Project Name: Lu Jing
Program Constructed by: Guangdong Xingyi Decoration Group
Designer: Ma Tingting

本案是三层的中式独栋别墅。负一层主要规划为休闲娱乐的空间，包括会客厅、书房、棋牌室，东北方向设计了一个景观露台，将自然的光影景致融入整个建筑。一层则规划为公共客餐厨区域，有两间卧室。客餐厅和卧室都设计了全开的推拉门，可直达庭院，最大限度地连接室内外空间。二层为睡眠区，有大小两个套间，主卧套间还带有独立的超大卫生间和步入式衣帽间，加长的露台展现了更宽阔的视野，使人与自然美景融为一体。

This project features a three-story standalone villa with traditional Chinese design elements. The basement is primarily designed as a leisure and entertainment space, including a reception hall, study, and game room. A scenic terrace on the northeast side integrates natural light and views into the architecture. The ground floor is planned as the public living, dining, and kitchen area, along with two bedrooms. The living, dining room and bedrooms have fully openable sliding doors, providing direct access to the courtyard and maximizing the connection between indoor and outdoor spaces. The second floor is designated as bedroom area, featuring two suites of different sizes. The master suite includes a spacious en-suite bathroom and a walk-in closet. An extended terrace offers expansive views, seamlessly blending the people with the natural beauty outside.

摩登品格
MODERN STYLE

项目名称：摩登品格
项目设计：广东星艺装饰集团
设计师：金嘉宜

Project Name: Modern Style
Program Constructed by: Guangdong Xingyi Decoration Group
Designer: Jin Jiayi

本案业主非常注重家居品质，通过深入沟通，风格确定为现代轻奢风，色调为暖咖色。本案原中空客餐厅、厨房在一楼，后将中空客厅、餐厅、厨房改至负一楼，与后面的花园同在一层，使人与环境和谐共存。

The homeowner of this project places a high priority on home quality. After in-depth discussions, the style was decided to be modern luxury with a warm coffee color palette. Originally, the high-ceiling living room, dining room, and kitchen areas were on the first floor. The designer later moved them to the basement level, aligning them with the garden at the back, creating a harmonious coexistence between people and their environment.

流金岁月

GOLDEN YEARS

项目名称：流金岁月
项目设计：广东星艺装饰集团
设计师：向白恒、吴中林

Project Name: Golden Years
Program Constructed by: Guangdong Xingyi Decoration Group
Designer: Xiang Baiheng, Wu Zhonglin

自然、空间、光影本是一体。客厅与自然相连，以一种平和的方式交相渗透，营造了静谧的空间氛围。沙发特有的柔白光感，让户外的“静”成为室内的“景”。餐厅简约的布局与客厅一脉相承。清雅气质的木饰面联结室内外的自然景色，宽阔明亮的落地窗与佳肴一同记录着晨昏日暮。

Nature, space, and light and shadow are inherently unified. The living room seamlessly connects with nature, blending harmoniously to create a tranquil atmosphere. The sofa's unique soft-white sheen transforms the outdoor serenity into an indoor scene. The dining room's minimalist layout echoes the living room's design. The elegant wooden finishes link the interior with the outdoor scenery, while the expansive floor-to-ceiling windows, alongside the delicious meals, capture the outside changing light of dawn and dusk.

李宅

LI'S RESIDENCE

项目名称：李宅
项目设计：广东星艺装饰集团
设计师：张志威

Project Name: Li's Residence
Program Constructed by: Guangdong Xingyi Decoration Group
Designer: Zhang Zhiwei

本案将优雅纳入空间，把精致融入生活。入户楼梯的屏风非常有趣，“隔而不断”这一思想在室内空间得以具体化。可旋转的屏风与圆柱相结合，增强了各个空间的联系，视野也得以延伸。客厅与餐厅、客厅与茶室互相对话。

为了凸显家居空间的精致美，本案使用金属、玻璃、大理石等光泽感和质感更好的材料，彰显高端品位。在空间颜色的搭配和主体色的选择上，除了黑、白、灰之外，本案还选择了驼色、浅咖色等自带高级感的颜色，材质则以古铜色金属不锈钢为主。

This project incorporates elegance and refinement into the space. The screen set at the entry is intriguing, embodying the concept of "separate but connected" within the interior. The combination of the rotatable screen and columns enhances the connectivity of various spaces, extending the visual field. The living room engages in a dialogue with both the dining area and the tea room.

To highlight the exquisite beauty of the home, the project employs materials such as metal, glass, and marble, which offer better luster and texture, reflecting a high-end taste. In terms of color coordination and primary color selection, besides the classic black, white, and gray, the designer also incorporates camel and light coffee tones, which reflect sophistication. Antique bronze stainless steel becomes the predominant material accordingly.

Grand House Scheme Design Works

4 大宅·方案设计作品

逐光而居——段宅

LIVING WITH LIGHT—DUAN'S RESIDENCE

项目名称：逐光而居——段宅
项目设计：广东星艺装饰集团
设计师：姚国健

Project Name: Living with Light—Duan's Residence
Program Constructed by: Guangdong Xingyi Decoration Group
Designer: Yao Guojian

本案是一套顶层复式，特殊的地理位置和视野环境给设计师带来不一样的心理感受。在设计规划时，设计师有意识地在所有空间引入大量的“光”，让主人感受到美好生活的“触手可及”。安藤忠雄曾说：“光赋予美以戏剧性，风和雨通过它们对人体的作用给生活增添色彩。建筑是一种媒介，使人们去感受自然的存在。”设计师把表面复杂的空间关系提升至建筑维度，研究光进入空间时的变化，将“逐光而居”的设计概念变为现实，发掘最核心的人与空间的需求关系，营造舒适的居住空间。

This project is a top-floor duplex. Its unique location and view provided the designer with a distinctive psychological experience. In the planning phase, the designer consciously introduced an abundance of "light" into all spaces, allowing the owner to feel that a beautiful life is "within reach". As Tadao Ando once said, "As light gives beauty its drama; or as wind and rain add color to life through their effects on the human body, architecture acts as a medium that allows people to feel the presence of nature." The designer elevated the complex spatial relationships to an architectural dimension, studying the changes of light that enters the space. The concept of "living with light" was hence realized, uncovering the fundamental relationship between people and space and creating a comfortable living environment.

随心而居

LIVING AS ONE WISHES

项目名称：随心而居
项目设计：广东星艺装饰集团
设计师：冯立龙

Project Name: Living as One Wishes
Program Constructed by: Guangdong Xingyi Decoration Group
Designer: Feng Lilong

思绪在城市的旷野中蔓延，游走在现实与幻想之间，感知自然与艺术，找寻生活的真意，让人与自然、艺术相联结，创造出生活的无限可能。

方正的布局，奢阔的尺度，客厅汇聚了家庭的温度，承载着生活的灵感。设计师巧妙地融入优雅与现代的元素，以灰咖色系和温馨的原木木纹，营造奢雅沉稳的质感。舒适的沙发、柔软的地毯和独特的艺术品相映成趣，简洁的线条和精致的细节为整个空间增添独特的格调。

Thoughts spread across the urban wilderness, wandering between reality and fantasy, sensing nature and art, and seeking the true meaning of life. Those thoughts, which connect people with nature and art, create infinite possibilities for living.

The square layout and spacious scale of the living room embody the warmth of family life and the inspiration of everyday moments. The designer skillfully integrates elegant and modern elements, using gray-coffee tones and warm-wood textures to create a luxurious yet serene ambiance. Comfortable sofas, soft carpets, and unique artworks complement each other, while clean lines and refined details add a distinctive style to the space.

16:21:09

餐厅是家庭共享美食的区域，本案以自然机理的米灰色涂料作为主色调，让原木色木饰面和石材桌面赋予餐厅氛围感，餐椅的绿意作为点睛之笔，为空间注入一丝明亮与清新的气息。一餐一食，感受生活的悠然与松弛，本案将功能与美学主义融合。主卧室是主人休息和放松的私密空间，浴缸区增加地台和灯带，巧妙地增加空间的层次感，打造出舒适而又迷人的卧室环境。柔软的灰咖色床品和精致的家具展现了高雅的品质，营造出恬静而又浪漫的氛围。在喧嚣的都市中找寻一方自然秘境，享受静谧与舒适，回归内心深处最本真的状态。

Dining area, as a place for family to enjoy meals, features natural textured beige paint as the main color. Wood veneers and stone tabletops enhance the atmosphere, with green dining chairs adding a touch of brightness and freshness. Each meal becomes a moment to savor life's ease and relaxation. This design merges functionality with aesthetics. The master bedroom, a private retreat for rest and relaxation, includes a raised platform and concealed lighting in the bathtub area, adding layers to the space and creating a comfortable, enchanting environment. Soft gray and coffee-toned bedding and exquisite furniture exude elegance, crafting a tranquil and romantic atmosphere. Amidst the bustling city, this space offers a hidden natural sanctuary where one can enjoy tranquility and comfort, returning to the most authentic state of inner peace.

在梦·草堂

IN THE DREAMY COTTAGE

项目名称：在梦·草堂
项目设计：广东星艺装饰集团
设计师：程炜

Project Name: In the Dreamy Cottage
Program Constructed by: Guangdong Xingyi Decoration Group
Designer: Cheng Wei

人总要沉下心来，喝茶读书，听雨看花，修篱种菊，热爱生活，感受家的温暖。

One always needs to calm down, enjoying daily affairs like drinking tea, reading books, listening to rain, admiring flowers, trimming hedges and planting chrysanthemums. People who love daily life can feel the warmth of home.

GONE GIRL

序
ORDER

项目名称：序
项目设计：广东星艺装饰集团
设计师：张大为

Project Name: Order
Program Constructed by: Guangdong Xingyi Decoration Group
Designer: Zhang Dawei

本案通过几个轴线的转场和延伸，达到类似园林对景的效果，使空间更具层次感。从狭长的过道到宽敞的室内空间，强烈的对比让人感受空间场景的开阔。在书房和餐厅的环形动线布局中，设计师考虑了使用的便利性和流畅度，使整个空间更加实用和舒适。窗台兼顾吧台的功能，在欣赏窗外美景的同时也可以放松休闲。

This project achieves a visual effect that the garden facing scenery through the transitions and extensions of several axial lines, enhancing the sense of layering within the space. The contrast from narrow corridors to spacious interiors creates a profound sense of openness. When planning the circular layout of the study and dining area, the designer has taken into account both convenience and fluidity of use, making the entire room more practical and comfortable. The windowsill also serves as a bar, providing a perfect spot for relaxation while enjoying the beautiful outdoor scenery.

一方素净

A SIMPLE AND TRANQUIL CORNER

项目名称：一方素净
项目设计：广东星艺装饰集团
设计师：范彪

Project Name: A Simple and Tranquil Corner
Program Constructed by: Guangdong Xingyi Decoration Group
Designer: Fan Biao

本案设计利用不同的曲线，给人带来不一样的视觉感受，搭配木色系的元素，呈现一幅美妙而灵动的画卷。独特的吊顶造型与独具个性的沙发展现出空间不一样的张力；搭配深色的吊式壁炉，既为空间补充元素，又为家增加温度。本案将饰面墙、电视柜、展示柜、收纳柜等进行有机串联，营造出连贯、轻简而自如的功能氛围。不同的饰面穿插其间，打造维度丰富、视觉饱满的一体化系统，更将意式风格的优雅与沉稳弥漫在空间的每个角落，业主在视觉、听觉得到满足的同时，心灵也得到慰藉。

This case utilizes various curves to create different visual experiences, together with elements in the wood color, presenting a wonderful and dynamic picture. The unique ceiling design and distinctive sofas bring out a different tension in the space. A dark suspended fireplace not only serves as a supplementary element but also adds warmth to the home. This project integrates the decorative wall, TV cabinet, display cabinets, and storage cabinets into a cohesive whole, creating a seamless and minimalist functional block. Different finishes are interwoven to create a rich visual experience. It further infuses every corner with the elegance and serenity of Italian style. While the homeowner's visual and auditory senses are satisfied, their spirit is also nourished.

WALTER FINANCE

山海居

RESIDENCE BY MOUNTAIN AND SEA

项目名称：山海居
项目设计：广东星艺装饰集团
设计师：钟华东

Project Name: Residence by Mountain and Sea
Program Constructed by: Guangdong Xingyi Decoration Group
Designer: Zhong Huadong

从都市到山间，从自然中觅得自在，揽漫天繁星，围垛夜谈，尘嚣暂抛，与自然共生，以原生态引领，由艺术感铺叙，设计师所思考的“和融”，基于人与人、人与自然、人与空间、人与生活的共契，不拘于特定范式，让自然时刻在场，让生活持续返场，自由而往，自在得趣，复归自我。

From urban to mountains, you may find tranquility in nature. Under the stars, gather around for night talks, and temporarily escape the hustle and bustle. Embrace coexistence with nature, guided by the essence of the original environment and express through an artistic narrative. The designer's interpretation of "harmony" is rooted in the connection between people, nature, space, and life. It is not confined to any specific paradigm. One should explore nature and life continuously. Freely relish in the ease, and return to oneself.

關雲長
BLADESMAN

保利君悦湾
POLY HYATT BAY

项目名称：保利君悦湾
项目设计：广东星艺装饰集团
设计师：许舰、周平秤

Project Name: Poly Hyatt Bay
Program Constructed by: Guangdong Xingyi Decoration Group
Designer: Xu Jian, Zhou Pingcheng

本案在设计时通过摆设家具建立了一个稳定的区域，满足日常需求。这样的空间构造能让家庭成员及访客在最佳位置欣赏花园风景。客厅选用耐用的木板来营造舒适的氛围。天花板和墙体则采用涂料与木材相结合的方式，与地面材料相协调。

The designer established a stable area by arranging furniturehas to meet daily needs. This spatial configuration allows people to enjoy the garden view from the optimal vantage points. The living room features durable wood flooring to create a cozy atmosphere. The ceiling and walls use paint and wooden materials to harmonize with the flooring materials.

春江明月

SPRING RIVER AND BRIGHT MOON

项目名称：春江明月
项目设计：广东星艺装饰集团
设计师：韦毅梅

Project Name: Spring River and Bright Moon
Program Constructed by: Guangdong Xingyi Decoration Group
Designer: Wei Yimei

逃离喧嚣，回归自然，亲近生活。面对快节奏的生活，能让人放慢脚步的生活空间才是最难能可贵的。本案改变了原户型的格局，利用自然灰色和胡桃木色的搭配，使空间更为静谧、稳重。

One should escape from the hustle and bustle, return to nature, embrace life. In today's fast-paced life, a living space that allows people to slow down is the most precious. This case has altered the original layout of the house, using a combination of natural grey and walnut colors to create a more tranquil and stable space.

山水之间

BETWEEN MOUNTAINS AND WATERS

项目名称：山水之间
项目设计：广东星艺装饰集团
设计师：王为

Project Name: Between Mountains and Waters
Program Constructed by: Guangdong Xingyi Decoration Group
Designer: Wang Wei

本案位于山水之间。设计追求无界而有情、无痕而有序。空间简约而有山石之趣，休闲而不失文化之韵。

This project is situated amidst mountains and water. The designer aims for boundless yet emotional, seamless yet orderly design. The area is minimalist with the charm of rocks, offering a relaxed atmosphere while retaining a cultural essence.

致简

SIMPLICITY

项目名称：致简
项目设计：广东星艺装饰集团
设计师：李佩耀

Project Name: Simplicity
Program Constructed by: Guangdong Xingyi Decoration Group
Designer: Li Peiyao

在动线设计上，本案以创造“游园”体验为底层逻辑；在拆除新建上，与“悬挑”“架空”“穿插”“片墙”“体块”等西方建筑的改造手法接近；在空间上，贯彻“少即是多”“眼前有景”的理念。

In terms of circulation design, this project is based on the concept of "garden tour". For demolition and reconstruction, it employs Western architectural techniques such as "cantilevering", "elevating", "interspersing", "isolated walls", and "modular designing". In terms of spatial design, it adheres to the principles of "less is more" and "views at every corner".

江东·印象

RIVER EAST IMAGES

项目名称：江东·印象
项目设计：广东星艺装饰集团
设计师：石金

Project Name: River East Images
Program Constructed by: Guangdong Xingyi Decoration Group
Designer: Shi Jin

本案户型为单边叠墅四层两套。在设计地下室时，设计师把本属于室内的空间借了一半作为车库，还将休闲区、家庭影院以及游戏区进行合理的划分，在家与朋友小聚、品酒品茶放松的同时，还能欣赏自己的爱车，何不乐哉！

This residence features two units of four-story, single-sided townhouses. For the basement design, the designer ingeniously allocated half of the indoor space for a garage, and thoughtfully divided the remaining area into a leisure zone, a home theater, and a game area. This layout allows owners to relax and enjoy wine or tea with friends while admiring their car—how delightful!

客厅拥享双层挑空，视觉观感与实际感受也随之开阔大气。相比之下，餐厅被刻意地“放低姿态”，时刻存在的烟火气也被“压缩”得更加浓厚。开阔的布局，只为让室内空间变得越发明亮通透。本案以全方位模式展开，把都市风采如数引入，添加了整体空间的层次美感。大客厅设计多种类几何家具，融合整体板块，直线、曲线互相呼应，营造出家的温馨亲切感。独特的造型配合巧妙的布局，尽显主人的“小心思”。墙体镂空，把空间利用做到最大化。底下摆置的篝火炉，燃烧着温暖的火焰，有一种家的温馨感。餐厅上方是一盏钢带状的吊灯，采用多层架构，赋予空间仪式感。整个用餐空间可坐下 6 人。岛台和厨房布局成吧台，必要时也可以当作临时餐厅或者酒吧。相比常规的餐厅，此处给人更多的舒适感。设计在改变居住环境的同时，也在潜移默化地改变生活方向与轨迹，设计承载的不仅仅是空间，更多的还是生活。

The living room features a double-height ceiling, creating an expansive and grand visual experience. In contrast, the dining area is intentionally designed to be at a lower elevation to help contain and minimize the spread of cooking fumes. The open layout makes the interior space more bright, open and airy. This project unfolds in a comprehensive manner, incorporating urban sophistication and adding a layered aesthetic to the whole space. The spacious living room features various types of geometric furniture, seamlessly integrated into the overall design. The interplay of straight and curved lines creates a warm and inviting home atmosphere. Unique shapes paired with a clever layout reveal the homeowner's thoughtful attention to detail. The hollowed-out walls maximize space utilization. A fire pit is placed below, with warm flames flickering, creating a cozy, homey atmosphere. Above the dining area, a steel belt-shaped chandelier with a multi-layer structure adds a sense of ceremony to the space. The entire dining area can accommodate six people. The island and kitchen are arranged like a bar, which can also serve as a temporary dining room or bar when needed. Compared to conventional dining rooms, this setup offers a greater sense of comfort. Design not only transforms living environments but also subtly alters the direction and trajectory of life itself. Design encompasses more than just space; it embodies life itself.

剪影
SILHOUETTE

项目名称：剪影
项目设计：广东星艺装饰集团
设计师：刘泽之

Project Name: Silhouette
Program Constructed by: Guangdong Xingyi Decoration Group
Designer: Liu Zezhi

多元素材质的组合，让空间更具可能性，从“新”出发，在“旧”的基础上创造属于“未来”的设计。灵感，在色彩之间跳跃，在几何之间迸发，在材质之间徘徊。

The combination of diverse materials infuses the space with greater possibilities, starting from the "new" and creating a design for the "future" on the basis of the "old". Inspiration leaps between colors, bursts from geometric shapes, and lingers among materials.

世纪

CENTURY

项目名称：世纪
项目设计：广东星艺装饰集团
设计师：赵寰

Project Name: Century
Program Constructed by: Guangdong Xingyi Decoration Group
Designer: Zhao Huan

本案是一座自建别墅，实用面积为 900 m²，是地块规划、建筑外观设计、室内规划与装饰设计一体化的项目。业主热爱收藏，对法式风格情有独钟，同时也希望融合中华传统文化，实现中西交融、和谐统一。

This project is a self-built villa with a practical area of 900 square meters, integrating site planning, architectural exterior design, interior planning, and decoration. The owner is a passionate collector, who favors French style and wants to integrate traditional Chinese culture to achieve a harmonious blend of Eastern and Western aesthetics.

黑金

BLACK GOLD

项目名称：黑金
项目设计：广东星艺装饰集团
设计师：褚万彦

Project Name: Black Gold
Program Constructed by: Guangdong Xingyi Decoration Group
Designer: Chu Wanyan

山本耀司说："黑色拥有谦虚与傲慢两种特质，黑色是慵懒随性却神秘莫测的，黑色是一种最有态度的颜色。"暗黑系是酷炫、个性、真实的代表，暗黑系风格没有浓烈的色彩分割，也不似灰色空间那样纯粹稳定，它展现出低调及从容的质感，以及暗黑的深邃。冷色调与暖光互相碰撞与摩擦，用最少的色彩创造出更丰富的层次。

Yohji Yamamoto said, "Black possesses two traits: humility and arrogance. It is casual yet mysteriously elusive. Black is the most assertive color." The dark color represents coolness, individuality, and authenticity. It lacks intense color divisions and does not offer pure stability like gray spaces; instead, it presents a subtle texture along with profound darkness. The collision and friction between cool tones and warm light create richer layers with minimal colors.

缘起

ORIGIN

项目名称：缘起
项目设计：广东星艺装饰集团
设计师：郭宇

Project Name: Origin
Program Constructed by: Guangdong Xingyi Decoration Group
Designer: Guo Yu

设计的目的是为生活中重要的事情创造一个最好的环境。人作为个体，永远是空间中最核心的元素。通过对空间感受和业主生活需求的分析，以及对比例、材料、光线的综合研究，本案将最简单的、最写意的空间设计呈现给业主。

The aim of design is to create the optimal environment for the most important aspects of life. People themselves are always the most core element in space design. Through analysis of the spatial perception and the homeowner's needs, as well as integrated studies on balance, materials, and lighting, this project presents the homeowner with the most minimalist and evocative space design.

五象印月

WUXIANG YINYUE

项目名称：五象印月
项目设计：广东星艺装饰集团
设计师：许舰

Project Name: Wuxiang Yinyue
Program Constructed by: Guangdong Xingyi Decoration Group
Designer: Xu Jian

对于家庭成员来说，客厅除了是多功能、多用途的空间，还是生活空间的中心。它的最佳位置是在餐厅旁边，与私人空间分开。本案在设计时，考虑了家庭成员人数、访客量和家族风格，通过摆设家具建立了一个稳定的区域，在所提供的空间里满足需求。

For family members, the living room serves not only as a multifunctional and versatile space but also as the center of their living environment. The best location of the living room is adjacent to the dining room, which separates from private space. Taking into account the number of family members, guest numbers, and family lifestyle, the designer has rearranged the furniture to create a stable area that meets needs.

Public Decoration Real-life Design Works

5 公共·工程实景作品

序·间
AMID THE ORDERS

项目名称：序·间
项目设计：广东星艺装饰集团
设计师：石全刚

Project Name: Amid the Orders
Program Constructed by: Guangdong Xingyi Decoration Group
Designer: Shi Quangang

本案是一个高定木作展厅，以“游园”的形式规划动线，通过“以小见大”的空间关系营造展示空间，让参观者在任意一个区域都能生发不同的情绪。

This project is a high-end custom woodworking showroom. It is designed with a "garden tour" concept to plan the circulation, and using "seeing the bigger through the small" as the spatial relationship to design the display areas. This approach allows visitors to experience different emotions in each area.

墅美美作
SPACEMAY

松庐和牛火锅餐厅

SONGLU WAGYU HOT POT

项目名称：松庐和牛火锅餐厅
项目设计：广东星艺装饰集团
设计师：叶子锋

Project Name: Songlu Wagyu Hot Pot
Program Constructed by: Guangdong Xingyi Decoration Group
Designer: Ye Zifeng

本案将《千里江山图》的色彩及元素贯穿到室内设计中，并与江月松庐相融合，在突出和牛料理就餐体验的同时，展现出时尚雅致的氛围。

This project incorporates the colors and elements of the *A Thousand Miles of Rivers and Mountains Painting* into the design, integrating them with the elements of river, moon, pine tree and cottage. While highlighting the dining experience of Wagyu cuisine, it also showcases a stylish and elegant atmosphere.

迭新与传承
EVOLUTION AND INHERITANCE

项目名称：迭新与传承
项目设计：广东星艺装饰集团
设计师：谢佳琪

Project Name: Evolution and Inheritance
Program Constructed by: Guangdong Xingyi Decoration Group
Designer: Xie Jiaqi

本案结合传统文化与现代气质，突破传统的办公模式，糅合自然、生活及艺术。整体空间以深色木纹为核心色调，现代流线与传统家具相融合，独树一帜的审美观与创新精神相得益彰。中式古典家具与现代设计家具融合一室，既形成了互补的视觉韵律，又展现出流畅的现代感，同时又不失文化的延续。

This case combines traditional culture with modern aesthetics, breaking the conventional office models by blending nature, life, and art. The overall space applies dark wood tones as the core color scheme, merging modern lines with traditional furniture. This unique aesthetic and innovative spirit complement each other perfectly. Classical Chinese furniture and modern design pieces coexist in the same room, showcasing a harmonious visual rhythm and modern aesthetic while simultaneously continuing cultural traditions.

小院烤肉
COURTYARD BARBECUE

项目名称：小院烤肉
项目设计：广东星艺装饰集团
设计师：洪泉

Project Name: Courtyard Barbecue
Program Constructed by: Guangdong Xingyi Decoration Group
Designer: Hong Quan

小院烤肉是一家融合户外和室内设计的现代简约风格餐厅。设计灵感来源于自然与现代的结合，旨在为顾客提供一个舒适、放松的用餐环境。入口处设置了一个小型庭院，种植了一些绿植和花卉。庭院的地面铺设了木质地板，与周围的绿植相互呼应，营造出一种自然的氛围。进入室内，墙壁选择了浅色调的中性色，如灰木色，使空间更加明亮。

Courtyard Barbecue is a modern minimalist style restaurant. Its outdoor and indoor areas are seamlessly integrated with the same design style. Drawing inspiration from the combination of nature and modernity, the restaurant aims to provide customers with a comfortable and relaxing dining environment. At the entrance, a small courtyard is adorned with greenery and flowers. The courtyard's wooden flooring complements the surrounding plants, creating a natural ambiance. Upon entering, you will see the walls are finished in light and neutral tones like gray-wood, making the space feel brighter and more inviting.

威龙酒庄

WEILONG WINERY

项目名称：威龙酒庄
项目设计：广东星艺装饰集团
设计师：鄢全富

Project Name: Weilong Winery
Program Constructed by: Guangdong Xingyi Decoration Group
Designer: Yan Quanfu

这座酒庄的室内设计充满了独特的魅力，为体验者提供令人难忘的品酒环境。无论是在树主题的酒窖中品尝美酒，还是在古典风格的展示接待区了解酒庄的历史，或是在宴会大厅举办各种活动，都能让人体验酒庄高品质的服务。

The interior design of this winery exudes unique charm, offering an unforgettable wine-tasting experience to its guests. Whether savoring fine wines in the tree-themed cellar, learning about the winery's history in the classic-style reception area, or hosting various events in the banquet hall, visitors are able to enjoy the winery's high-quality services.

国信·宜和 / 一多科技

GUOXIN YIHE/YIDUO TECHNOLOGY

项目名称：国信·宜和 / 一多科技
项目设计：广东星艺装饰集团
设计师：柯君

Project Name: Guoxin Yihe/Yiduo Technology
Program Constructed by: Guangdong Xingyi Decoration Group
Designer: Ke Jun

本案为空调、地暖、新风、智能、灯光一体化的智能展厅体验店。

商业空间设计的目的是给客户提供体验空间，展示商品及其功能，促进消费，引领新的消费及生活方式。设计师从各个方面入手，直观地体现其展示性、服务性、促销性和文化性。该展厅共两层，面积共 468 ㎡，主要由展示区与休息区两部分组成。展厅基调为暖色，以浅黄色为主。展区以深色金属和原木制成，接待座椅以现代风格为主，地面铺设为灰色和暖色，吊顶为白色，墙壁为暖色。分区一楼为冷色，主要体现科技感；二楼为暖色，产品以家装为主，体现了家的温馨。整个空间富有层次感，提升了展示的效果。

This project is an intelligent exhibition and experience store integrating air conditioning, floor heating, fresh air, intelligent systems, and lighting systems.

The purpose of design is to provide customers with an experiential space, showcasing products and their functions, promoting consumption, and leading new consumption and lifestyle trends. Designer intuitively presents its function of display, service, promotion and culture. This exhibition hall has two floors, with a total area of 468 square meters, mainly consisting of a display area and a rest area. Its overall tone is warm, primarily featuring light yellow. The display area is constructed from dark metal and natural wood, with modern-style reception seating. It features gray or warm-toned flooring, white ceiling and warm-toned walls. The first floor has a cool color scheme, mainly reflecting a sense of technology. The second floor has a warm color scheme, as the products presented are primarily focused on home decoration, reflecting the warmth of home. The entire space is rich in layers, enhancing the overall display effect.

COLMO
高定纯嵌系列
烹饪体验中心
COLMO

COLMO
COLMO

HEALTH
NEW LIFE STYLE

全球超级个体COLMO非凡人生
COLMO

Public Decoration Scheme Works

6 公共·方案设计作品

新翼

NEW WINGS

项目名称：新翼
项目设计：广东星艺装饰集团
设计师：郭晓华、陈文辉、廖文涛

Project Name: New Wings
Program Constructed by: Guangdong Xingyi Decoration Group
Designer: Guo Xiaohua, Chen Wenhui, Liao Wentao

承于艺术，融于时代，三十二年集涓成河，形滔滔之势，星艺装饰抒写着自己的故事，一页页篇章下的历史和记忆沉淀后如星辰般闪耀，指引着星艺汇金全案中心开启新的航程。

Rooted in art and merged with the times, history of thirty-two years gathered into a mighty river, embodying a powerful momentum. Xingyi Decoration Group has been crafting its story for thirty-two years; each chapter, enriched by history and memory, shines like stars. This guides the Xingyi Decoration Group Huijin Full Case Center into a new journey.

前区的水晶玻璃砖墙共有 3 232 块，长达 13.2 米，正如星艺三十二载专注筑家，循梦而行。当然，不止于此，几代星艺人高举坚守的火种，用炽热的火苗点亮未来，不论时代如何更迭变迁，星艺都会在光亮下延绵生长，永不止息。

水晶玻璃砖墙朦胧通透的质感正如星艺全案设计，为每位客户提供公开、公正、透明的一体式装饰服务和家居解决方案，让追求个性化的业主真正体验定制化的设计，实现让客户省心、省钱、省时的愿望。以框景手法将背后的木棉花艺术画显露一角， 让人忍不住想一探究竟。

The front area's crystal glass brick wall consists of 3,232 pieces, stretching 13.2 meters, symbolizing Xingyi Decoration Group's thirty-two years of dedication to building homes and pursuing dreams. Surely that's not all. Generations of Xingyi Decoration Group's employees have upheld the flame of commitment, using its burning intensity to light up the future. Regardless of how the times change, Xingyi Decoration Group will continue to grow under the light, never ceasing.

The crystal glass brick wall's hazy and translucent texture is like the comprehensive design of Xingyi Decoration Group, offering clients open, fair, and transparent integrated decoration services and home solutions. This allows owners pursuing personalization to truly experience customized design, fulfilling the wish for convenience, cost savings, and time efficiency. The framing technique reveals a glimpse of the kapok flower painting behind, tempting one to explore further.

史丹利定制展厅

STANLEY CUSTOMIZED HALL

项目名称：史丹利定制展厅
项目设计：广东星艺装饰集团
设计师：范建国

Project Name: Stanley Customized Hall
Program Constructed by: Guangdong Xingyi Decoration Group
Designer: Fan Jianguo

多代同堂的传统四合院是儒家思想在栖居空间上的诠释，“同一屋檐”隐喻了共居概念。在展厅项目设计中，设计师从四合院中汲取灵感，在中部空间中引入了天井的概念，使各个功能展示空间在屋檐下如同庭院的各个厢房一样融合在一起。客人游走其中，就像在朋友家串门一样自在和放松。

整个空间的规划既满足了功能性，也表达了栖居者的情感依恋。同一屋檐下的意象与四合院的建筑形式相融合，为展厅空间带来新的诠释。

The traditional si he yuan (courtyard house) with multiple generations living under the same roof is an embodiment of Confucian thought in residential spaces. The "same roof" metaphorically represents the concept of living together. In the design of this exhibition hall, the designer drew inspiration from the courtyard house, incorporating the concept of a central courtyard. This design approach integrates the various functional display areas under the same roof, much like the way that individual rooms in a courtyard house are interconnected. Guests wandering within the space experience a sense of ease and relaxation, akin to visiting a friend's home.

The spatial planning not only meets functional requirements but also expresses the emotional attachment of its residents. The imagery of "under the same roof" blends with the architectural form of the courtyard house, bringing a fresh interpretation of the exhibition hall.

HELLO
SUMMER

半壁茶事

HALF-WALL TEA HOUSE

项目名称：半壁茶事
项目设计：广东星艺装饰集团
设计师：冯立龙

Project Name: Half-Wall Tea House
Program Constructed by: Guangdong Xingyi Decoration Group
Designer: Feng Lilong

“朴隐山居待雅士，静修茶事候知音”，这是半壁茶事创始人一直坚守的初心。

设计师也遵从这个起点来规划这个茶空间。入口处暗调的光影折射在水面上，柔和的金色光影引导着前行的步伐，与水面上的花艺景观融为一体。设计师以当代的克制手法呈现东方美学精神，营造能够与人和平对话的茶空间形态，使人完全沉浸于宁静的茶意世界中。松下品茗赏月看云烟，虚实相生。一束光仿佛可以通过空间照射进人的内心。跟随着光的引导步入室内，空间体验缓缓展开。

"Waiting for a cultured gentlemen in a humble cottage, quietly preparing tea for a kindred spirit" — this is the founding principle that the creator of Half-Wall Tea House has been adhered to.

The designer follows this principle in planning this tea house. The dimly lit shadows at the entrance are reflected gently on the water's surface, with soft golden hues guiding guests step forward. The light and shadow seamlessly blend with the floral decoration on the water. Using a contemporary restrained approach, the designer presents the spirit of Eastern aesthetics, creating a tea house that fosters a peaceful dialogue with its visitors, allowing them to fully immerse themselves in a tranquil world of tea. In this tea house, you can enjoy an experience: tasting tea while gazing at the moon and drifting clouds; reality and imagination seamlessly blend. A beam of light seems to penetrate through the space and into the soul. Following the light, one can experience this space gradually.

南麓售楼部

SALES CENTER OF NANLU

项目名称：南麓售楼部
项目设计：广东星艺装饰集团
设计师：陈华、罗骁

Project Name: Sales Center of Nanlu
Program Constructed by: Guangdong Xingyi Decoration Group
Designer: Chen Hua, Luo Xiao

南麓售楼部处处都体现了企业文化和时代潮流的结合，表现出强烈的新古典主义色彩。“使宾客得到最大限度的满足，得到最优质的服务，从精神上到物质上。”这是售楼部设计的宗旨。售楼部设计突出深厚的文化底蕴，尽力满足销售中心的文化追求和中高端住宅客户的目标追求。因此，设计不应囿于形式上的创新，更应进行深层次挖掘，以充分的精神内涵赋予建筑产品以鲜活的生命，在喧嚣的尘世中创造出“高树蝉声秋巷里，朱门冷静似闲居”的无限意境。

The Sales Center of Nanlu embodies the combination of company culture and contemporary trends, exhibiting a strong neoclassical style. "To provide guests with the utmost satisfaction and the best service, both spiritually and materially." —This is the design purpose of the sales office. The design of the sales center highlights a profound cultural heritage, striving to meet the cultural pursuits of the estate company and the target pursuits of residential clients. Therefore, the design should not be limited to formal innovation but should delve deeper to impart vibrant life to the architectural design through ample spiritual connotation. This is to create infinite artistic conception in the bustling world, akin to "Early autumn, cicadas chirp in the tall trees in the alley; the noble mansion is serene as a leisurely residence".

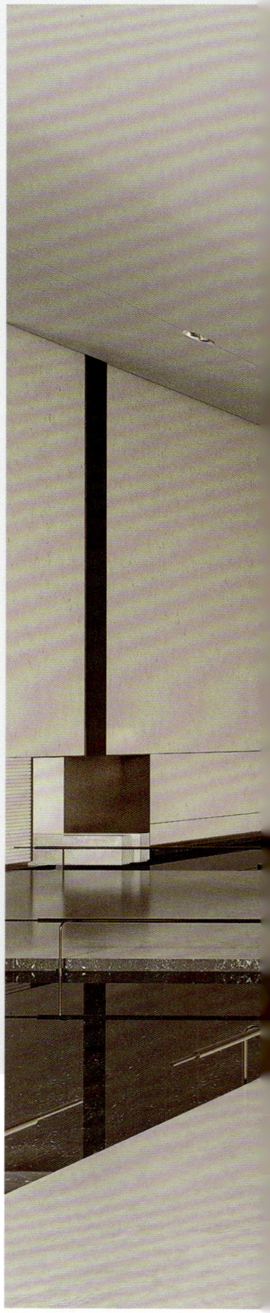

展厅空间

EXHIBITION HALL

项目名称：展厅空间
项目设计：广东星艺装饰集团
设计师：张大为

Project Name: Exhibition Hall
Program Constructed by: Guangdong Xingyi Decoration Group
Designer: Zhang Dawei

展厅设计以游走路线为基础，将展示区与动线紧密结合，创设了一个开放、通透的空间。平面布局灵活自由，空间秩序层次分明。明亮的色调和巧妙的灯光照明，能够突出展示重点，达到最佳的展示效果。展品的陈列设计注重给客户留下深刻的印象。

The exhibition hall designing integrates the display area and people's circulation, creating an open and transparent space with flexible layout and clear spatial order. Bright colors and ingenious lighting highlight the main displaying objects and achieve best effect, leaving deep impression on customers.

湘未院子

XIANGWEI YARD

项目名称：湘未院子
项目设计：广东星艺装饰集团
设计师：洪泉

Project Name: Xiangwei Yard
Program Constructed by: Guangdong Xingyi Decoration Group
Designer: Hong Quan

本案餐厅以湘菜为特色，位于一个占地 1 500 m^2的院子内，院内设有庭子、散座和水池。

院子是餐厅的亮点之一，为客人提供了舒适、自然的用餐环境。庭子的设计采用传统的江南园林风格，以石头、花草和小径为主要元素，营造出宁静和谐的氛围。散座区域则布置有舒适的户外桌椅，供客人在户外用餐或休闲。水池的设置增添了一丝清凉感，为整个院子增添了一抹生机。

This case is a restaurant, primarily serving Hunan cuisine. It is situated in a 1 500-square-meter courtyard that features pavilions, separated seats, and a pond.

The courtyard is one of the restaurant's highlights, offering guests a comfortable and natural dining environment. Designed in the traditional Jiangnan-garden style, the courtyard incorporates elements such as stones, flowers and winding paths to create a serene and harmonious atmosphere. The separated seat area is furnished with cozy outdoor tables and chairs, providing a relaxed setting for dining or leisure. The pond brings a refreshing feeling and adds a lively element to the entire courtyard.

川蜀印象

SICHUAN IMAGE

项目名称：川蜀印象
项目设计：广东星艺装饰集团
设计师：郭宇

Project Name: Sichuan Image
Program Constructed by: Guangdong Xingyi Decoration Group
Designer: Guo Yu

传统川菜文化在与现代时尚空间的汇归中，促成了微妙且平衡的适切状态。本案采用意象手法，透过毛玻璃、金属框架、木与石等材质，凸显层出叠见的自然气度。所有概括、凝练的具象营造均顺从人的内心。立身街面，我们希望透过一层空间溢散的柔和灯光，创设诚挚迎接每位食客的意境。疏影为序，光影朦胧缔造感知链接。简洁的线条经毛玻璃散射，渲染洗光效果，划分区域却保持整体气氛的通融。

The traditional Sichuan cuisine culture merges with modern spaces, creating a subtle and appropriate balance. This project adopts impressionistic techniques, using frosted glass, metal frames, and materials like wood and stone to highlight the emerging and overlapping natural views. All summarized and refined constructions align with the human heart. Standing on the street, we hope that the light emitting from the first floor can create an inviting atmosphere, sincerely welcoming every guest. Shadows set the tone, while the interplay of light and shadow fosters a sensory connection. Simple lines diffuse through frosted glass, creating a light effect that divides areas while maintaining a cohesive ambiance.

海威迈信

HEAVSTY

项目名称：海威迈信
项目设计：广东星艺装饰集团
设计师：柯君

Project Name: Heavsty
Program Constructed by: Guangdong Xingyi Decoration Group
Designer: Ke Jun

本案旨在为划线机展厅提供一个展示空间，以展示划线机品牌形象、机型特点和售后服务，吸引更多客户进店选购。通过合理的空间规划、灯光设计、色彩搭配以及智能化系统的应用，提升客户购车体验和售后服务质量，为划线机企业带来更多的商业机会。

This case is an exhibition hall for line-marking machines, showcasing Heavsty's brand image, machine features, and after-sales service. It aims at attracting more customers to visit and make purchases. By implementing effective space planning, lighting design, color coordination, and smart systems, this exhibition hall enhances the customer experience and after-sales service quality, while also creating more business opportunities for the company.

刘宅

LIU'S RESIDENCE

项目名称：刘宅
项目设计：广东星艺装饰集团
设计师：刘文达

Project Name: Liu's Residence
Program Constructed by: Guangdong Xingyi Decoration Group
Designer: Liu Wenda

自然质朴的材质用当代的手法重叠构建，由高处洒落的自然光，与耳边若有似无的流水声，让这段旅程的编排，如探山间，身临其境。有山便有门，有门便有路。探于“山间”，路与门随着景色与光影而不断延伸，创造不同的试探性选择路径，行者可自由地停留或在迂回中前进。所有墙体看似隔绝，却也能让行者开拓出不同的“登山”路径。游走其间，留下探索的乐趣，稍显压抑的前景为旅程后期的豁然埋设情绪的伏笔。每个步入其中的人，或许都将有不同的感受，不被定义的体验，是设计师希望传达的一种对“自由”的思考。

Natural and simple materials are constructed with contemporary techniques. The natural light cascading from above, combined with the subtle sound of flowing water, creates an immersive experience akin to exploring a mountainous terrain. Where there is a mountain, there is a door; and where there is a door, there is a path. In the "mountainous terrain" exploration, paths and doors extend with the changing scenery and light, creating various exploratory routes. Travelers can freely pause or navigate through meandering pathways. Though the walls may seem isolating, they also allow for the discovery of different "mountain climbing" routes. Wandering through this space leaves room for exploration and enjoyment, with the somewhat oppressive foreground setting the stage for the emotional revelations to come later in the journey. Every person who steps into the space may have different experiences. The undefined nature of this experience reflects the designer's contemplation of "freedom".

慢居
SLOW LIVING

项目名称：慢居
项目设计：广东星艺装饰集团
设计师：刘树东

Project Name: Slow Living
Program Constructed by: Guangdong Xingyi Decoration Group
Designer: Liu Shudong

本案属于民宿设计，利用村民原有的闲置空房及可使用的空地来规划设计，有效地依托和利用周边环境与资源，使人与自然和谐相适，同时利用原结构可利用的资源，结合新规划的结构来营造独特的生活意境。装修的特点是淡雅、简洁，空间线条也比较清晰，使居室的布置给人以清洁、舒适的感觉，让人充分体验“慢生活”。色彩偏重于原木色，形成朴素的自然风格。

This design is for a homestay, which utilizes the villagers' existing idle rooms and available land. It effectively utilizes the surrounding environment and resources to harmonize people with nature. By fully making use of the original structure and planning new structure, the designer creates a unique living ambiance. The renovation features a subtle and simple design, with clean lines that give the space a fresh and comfortable feel, allowing guests to fully experience a "slow living" lifestyle. The color scheme emphasizes natural wood tones, creating a rustic and understated natural style.

muse 缪斯

MUSE

项目名称：muse 缪斯
项目设计：广东星艺装饰集团
设计师：李莎

Program Name: Muse
Program Constructed by: Guangdong Xingyi Decoration Group
Designer: Li Sha

本案一楼是大厅，为接待和洽谈区域。大厅采用明亮的自然光线和简洁的装饰，营造出宽敞明亮的氛围。接待台采用简约的设计，使用一体成型的微水泥涂料，以展现高端感。洽谈区域则为顾客进行详细的咨询和讨论提供了空间。

二楼是美容包间。每个包间都有足够的空间供美容师和顾客进行交流。包间内的装饰简约而不失温馨，使用柔和的灯光和舒适的家具，如按摩床、梳妆台和储物柜。

The first floor of this case is the lobby, an area for reception and negotiation. The lobby utilizes bright natural light and simple decorations to create a spacious and bright atmosphere. Meanwhile, the reception desk is designed with minimalism style, using seamless microcement coating to convey a premium sense. And the negotiation area provides customers with broad space for further consultations and discussions.

The second-floor houses private beauty rooms, each offering enough space for interaction between the beauticians and their clients. The decoration within these rooms is simple yet cozy, featuring soft lighting and comfortable furniture such as massage beds, dressing tables, and storage cabinets.

AM
10 00

宴·山江荟

FEAST OF MOUNTAINS AND RIVERS

项目名称：宴·山江荟
项目设计：广东星艺装饰集团
设计师：屈静月

Project Name: Feast of Mountains and Rivers
Program Constructed by: Guangdong Xingyi Decoration Group
Designer: Qu Jingyue

本案是私厨餐饮空间，在整体的空间布局中，考虑交通流线，将空间分为不同区域，包括前台、包间、大厅、厨房等。本案将自然元素融入餐厅空间，以营造舒适和放松的氛围，使用木材、石材和绿植等营造出自然的触感和视觉效果。整体空间以黑色和木色饰面天花板，以及木色护墙板、红色纹样护墙等作为装饰，配合局部造景手法，融入绿植、流水、岩石等元素。因为原空间整体的采光不是很充足，所以本案将整体沿楼梯部分的墙体做了透光板处理，以增加空间的采光。

This case is designed for a private restaurant. In planning the overall layout, the designer considered circulation and divided the space into different areas, including a reception area, private rooms, a main hall, and a kitchen. Natural elements are integrated into the restaurant to create a comfortable and relaxing atmosphere, using materials such as wood, stone, and greenery to evoke a natural feel and visual appeal. The overall space features black and wood-tone decorated ceiling and utilizes wood-tone and red patterned wainscoting as decorations. Elements like greenery, flowing water, and rocks are applied in local landscaping. Due to the original space lacks of sufficient light, translucent panels were installed along the walls near the stairs to enhance the natural light.

201
202

中科采象

EVERACQ TECHNOLOGIES

项目名称：中科采象
项目设计：广东星艺装饰集团
设计师：辛建华

Project Name: Everacq Technologies
Program Constructed by: Guangdong Xingyi Decoration Group
Designer: Xin Jianhua

本案是一家科技研发类公司，位于中国长三角城市群副中心城市，是国家重要科研教育基地。设计师深谙美学意境，打造了敞阔的现代办公空间。整体设计充满建筑美感，空间布局合理，打破传统的办公规划流线，实现灵活布局，休闲区与公共办公区相连接，造就高效、快捷的办公环境。办公室拥有充足的自然光线和极佳的视野，温润的木质感与坚固的架构造型相结合，呈现出超凡的空间气度。

This project is for a technology company located in Hefei, a secondary central city of China's Yangtze River Delta region. Hefei is also known as an important base for China's research and education. The designer, with a deep understanding of aesthetic concepts, created a spacious modern office space. The overall design is architecturally elegant with a well-organized layout that breaks model of traditional office traffic flows, achieving a flexible arrangement. The leisure area is connected to the public office area, creating an efficient and fast work environment. The office features ample natural light and excellent views, with a design that combines the warmth of wood textures and solid structural shapes, presenting an extraordinary sense.

设计幸福

耕艺种德